LIGHT, MOLECULAR MECHANISM & SLEEP (BASICS)

LIGHT IS THE GOVERNOR OF THE UNIVERSE

GIRISH BHARDWAJ

INDIA • SINGAPORE • MALAYSIA

Notion Press Media Pvt Ltd

No. 50, Chettiyar Agaram Main Road,
Vanagaram, Chennai, Tamil Nadu – 600 095

First Published by Notion Press 2021
Copyright © Girish Bhardwaj 2021
All Rights Reserved.

ISBN 978-1-63904-769-7

This book has been published with all efforts taken to make the material error-free after the consent of the author. However, the author and the publisher do not assume and hereby disclaim any liability to any party for any loss, damage, or disruption caused by errors or omissions, whether such errors or omissions result from negligence, accident, or any other cause.

While every effort has been made to avoid any mistake or omission, this publication is being sold on the condition and understanding that neither the author nor the publishers or printers would be liable in any manner to any person by reason of any mistake or omission in this publication or for any action taken or omitted to be taken or advice rendered or accepted on the basis of this work. For any defect in printing or binding the publishers will be liable only to replace the defective copy by another copy of this work then available.

LIFE DID NOT BEGIN WITH MY BIRTH NOR SHALL IT END
WITH MY DEPARTURE…

ALL THAT MATTER IS, MY CONTRIBUTION BEFORE MY
DEPARTURE…

CONTENTS

Inquisitiveness & deep rooted convictions 7
1. Introduction 11
2. Light & Circadian 17
3. Physiology & Light.. 31
4. An Insight: The solutions 39
5. Light Therapy.. 49
6. Sleep 61

Conclusion 67
References and credits 69
About Theralicht™ LLP 71

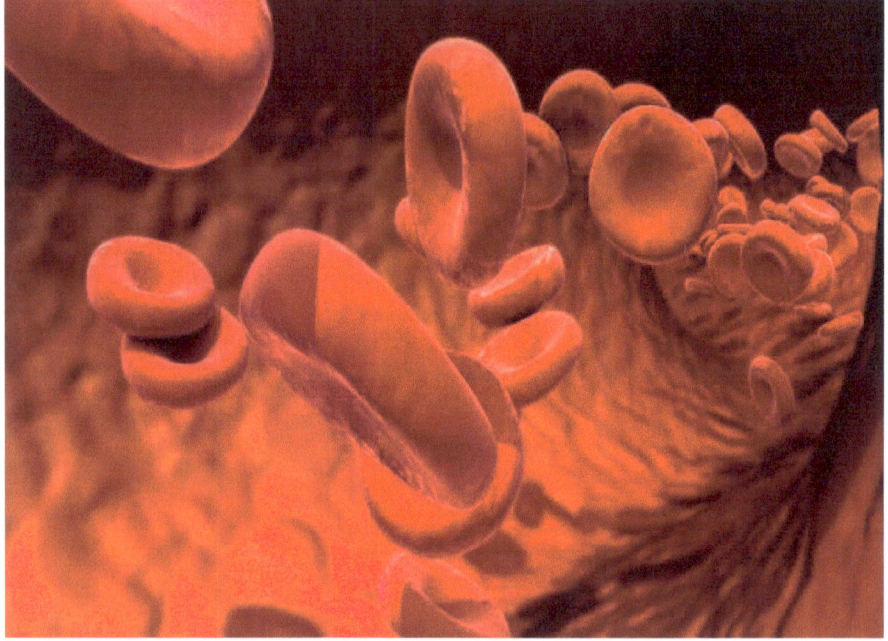

INQUISITIVENESS & DEEP ROOTED CONVICTIONS

....

The marathon of my journey in the field of lighting started in the year 2000.

The real understanding on the needs of light rather the subject LIGHT started from the year 2007. I started searching for the books, people and companies who would have made me aware of the word LIGHT.

Then suddenly one fine morning during the dawn (while I was on the window seat of the aircraft up in the sky), I observed the Sun (as if I have never seen a sun before) and that jiffy was the beginning of the self-study.

I finally could get some answers but still was not able to understand in a very explicit way, that how the phenomenon works as there were many preposterous but some relevant questions within myself.

{Started studying some subjects like physics, geothermal, material sciences and a bit of electrical sciences.}

Finally in the year 2012, I wandered thru some relevant answers and started working around it. In the year 2014 my search soared to the new horizons on LED and LIGHTING.

By the time I was getting aware of LED and LIGHTING, in the year 2015, I started studying some parameters of human body, molecular mechanisms, the fundamentals of our body genes and plants so that finally I can understand how light has governed them.

There is the inferred assumption that as humans we are unencumbered by the demands levied by our biology and that we can do what we want, at whatever time & the ways we choose, even after experiencing that more than 70% of our time is spent in the indoors.

I want to extend my services in spreading awareness about this for the betterment of humankind and to begin with this, let there be two way interaction through this book of mine.

We Humans and all the living being are already a part of the cosmic world and the forces. Especially referring to our planet with it's various dynamical changes of tilt, rotation, and revolution on the eccentric orbital path & our desired living needs.

Every thought, feeling, and emotion we experience is actually triggered by specific vibrations that resonate on the grid of our energetic matrix.

This would imply that specific frequencies of energy or light (color), trigger explicit thoughts, feelings, and emotions within us in the same way that certain life experiences do.

Thus, how we interact with the spectrum of light affects our mind and body, as well as being inseparably linked to how we relate to the full spectrum of life's experiences. Just as some life experiences are easy and others are difficult, some colors are comfortable to look at, while others feel uncomfortable. Because of this discomfort, our bodies respond to light *selectively*, absorbing certain colors and rejecting others.

Since every aspect of life is light dependent, rejecting any portion of the visible spectrum decreases the flow of our life force, diminishing our innate healing ability and our capacity to experience health and happiness.

However, we can magnify our ability to receive the visible and non-visible parameters of light to correct the physiological imbalances associated with them.

As goes unmentioned, that the **"light is the governor of the universe©"**, it is important to understand the physiology, socio-psychological & retino-hypothalamic aspects being governed by lighting parameters like frequency of SPD, energy eV impacts and others.

It's our responsibility to initiate the sprouting process through the respective parts of the brain for better cognitive skills, limbic experience and assist the neurotransmitters

The human energy system is in a continual ever changing & in holistic state. This harmony is accomplished by constantly synchronizing our body's atoms, glands, genes, molecules, cells hormones in alignment of time cues- the LIGHT.

Acting as the body's light meter, the pineal concurrently communicates information about the time of day, time of year, spectral characteristics, and the earth's electromagnetic field with every cell in the body *simultaneously*. For instance the protein encoded by the CLOCK gene plays a central role in the regulation of these circadian rhythms via its function as a transcriptional regulator of genes involved in maintaining sleep/wake cycles

In so doing, each cell arranges its internal function and synchronizes itself with Mother Nature.

Whereas in reality every aspect of our physiology and behavior is constrained by a 24 h beat arising from deep within our evolution.

The combined and interlocking effects of physiological, psychology and psychosocial stress lead to emotional, cognitive, and physiological pathologies.

The idea that there can be an "explicit technique" for human-bio favorable lighting that can be universally applied to all persons is unrealistic, because of the tremendous variability between individuals based on their age, genetics, environmental factors, individual physiology, eating habits, and wake up & sleep timings and personal behaviors.

On the other hand plants, honey bee, bats and many other living doesn't respond to the visible spectrum as we do. It perceives reality within the ultraviolet portion of the spectrum same with the bats, while a snake interprets reality through the infrared portion.

Chapter 1

INTRODUCTION

Now I would like to leave all those readers triggering their inquisitiveness regarding the interesting facts of light. I need you to search for the answers, if you are aware then applauds, if you are unaware even then applauds as you will definitely get awareness once you are in search of the answers for these questions:

1. Why Sky appears to be blue?
2. Why sky, in the dusk and dawn hours appears to be red, amber or golden?
3. What makes northern light so interesting?
4. Why sunlight is the best medicine? (With recommended exposure time and place)
5. Why we feel hot in the sun?
6. How 95% of the invisible light acts as the governor of hormones and universe?
7. What is that molecular mechanisms in the plants which makes them more intelligent than us when it comes to the phenomenon of using the best of the light. (Here I am not referring to photosynthesis)

You can Google it out, you can search by yourselves going thru some references or the books or you may ask the correct person from lighting or related domain (can be a teacher, researcher, scientists, and enthusiast … on and on…)

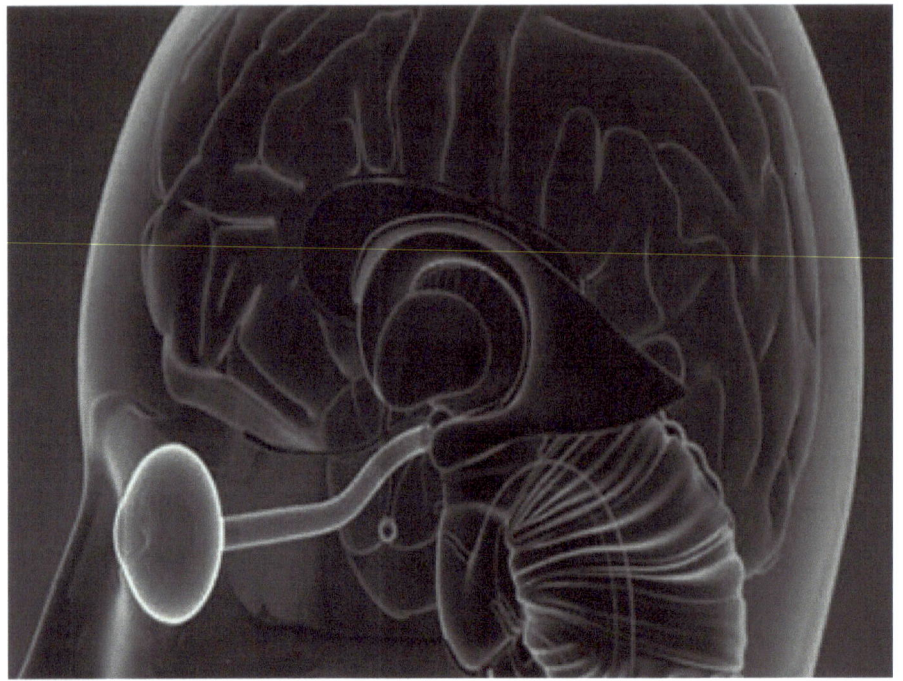

VISUAL & NON VISUAL IMPACT OF LIGHT

1. Retinal damage
2. Brain fatigue
3. Stress
4. Migraine
5. EFD
6. Obesity
7. Weakened skeletal muscle
8. Age related macular disorder
9. Prone to cancer

Before you go further reading down the book, I would like you all to recollect and hold on to one fact that the DNA of LIGHT has changed with the arrival of LEDs. Previously light was in the form of heat energy but now it's not.

Thus if the DNA has changed so do our way of approach on managing the parameters of light for the best of us all has to be in line, and for that we have to struggle our brains to offer the best to the users (like office goers, residences, hospitality & health care).

The other hidden fact is that only 1/3rd part of the energy in our body comes from the food rest 2/3rd comes from the radiations and the light.

- **What is Light?**

Visible light is electromagnetic radiation within the portion of the electromagnetic spectrum that can be perceived by the human eye.

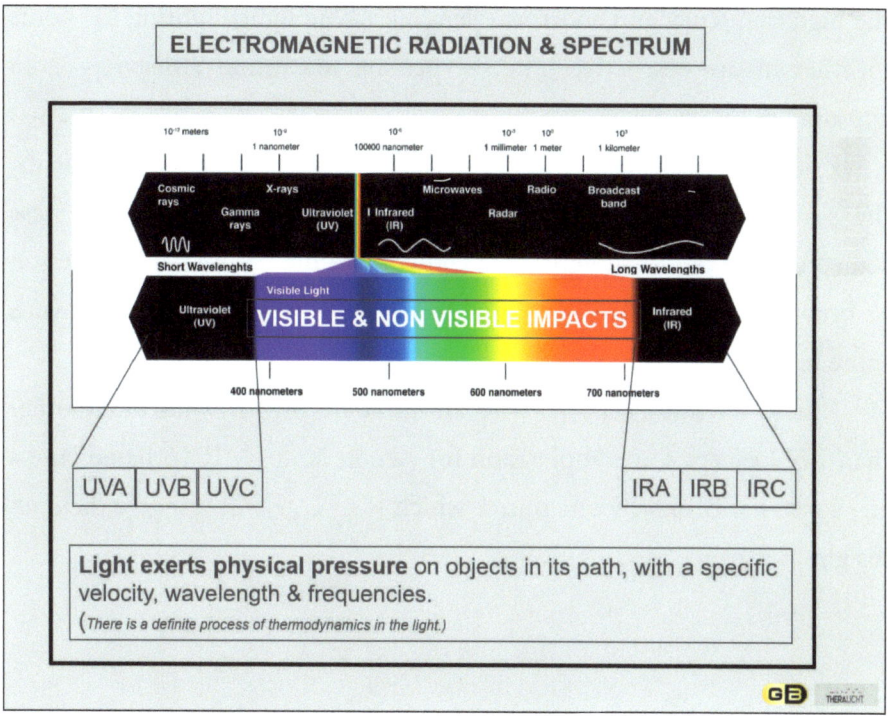

Visible light is usually defined as having wavelengths in the range of 400–700 nanometers (nm).

Light sometimes refers to electromagnetic radiation of any wavelength, whether visible or not. In this sense, gamma rays, X-rays, microwaves and radio waves are also light.

Light exerts physical pressure on objects in its path, of velocity, wavelength & frequencies. (*There is a definite process of thermodynamics in the light.*)

The photon is a type of elementary particle. It is the quantum of the electromagnetic field including electromagnetic radiation such as light and radio waves, and the force carrier for the electromagnetic force.

In today's world, the nights are days and the days are nights for many of us.

We are exposed under one type of Static/dynamic LED (which have the high frequency and short wavelength- a blue light) lighting for hours together in our respective offices expecting maximum efficiency & an appropriate behaviour.

I do not want to get into cliché that what is LED, the components, the parameters, what is lux, what is lumen etc etc…however in case someone wants to learn light, they can contact me or any lighting person or browse thru the internet or get good books or enrol themselves with some institutions.

Just to let those readers, who are unaware on LED (as today other than LED except some application for Xenon lamp, MR 16 based lamps etc) Light has primarily one source which is LED, I will just give them an insight at a glance.

Introduction ◆ 15

- **LED, Semiconductors and wavelengths**

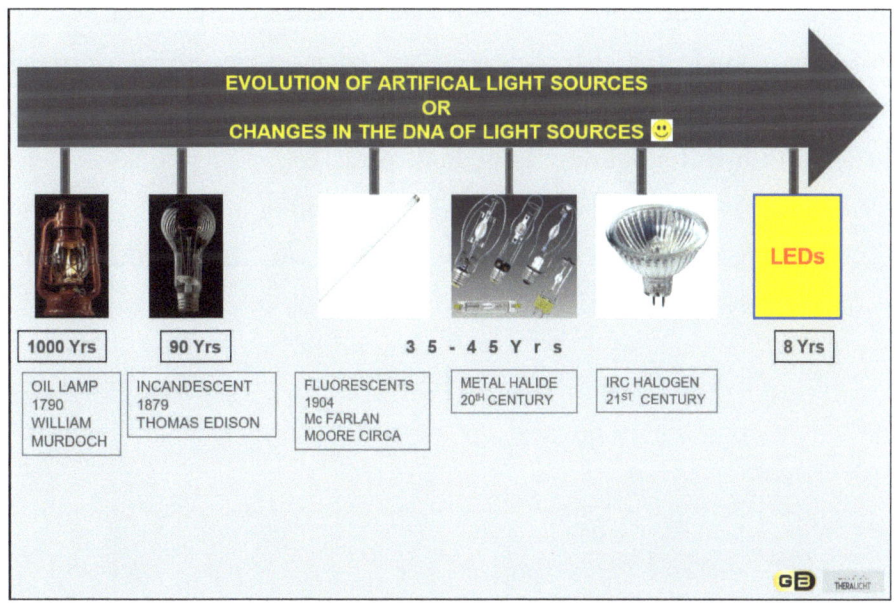

LED, Semiconductors and wavelengths

A semiconductor that radiates light when subjected to electrical impulses – a phenomenon called electroluminescence.

An acronym for light-emitting diode is LED,

LED runs on direct current (DC) and often requires a separate electrical ballast – a 'driver'. The driver converts the mains voltage to an optimal level for the LED.

Sl.No	Semiconductor material	Wavelength nm	Color	Vf@20mA	Ref applications
1	GaAs	850-940	iR	1.2	TV remotes etc
2	GaAsP	630-660	Red	1.8	Indicating lights
3	GaAsP	605-620	Amber	2.0	Indicating lights
4	GaAsP:N	585-595	Yellow	2.2	Indicating brighter lights
5	AlGaP	550-570	Green	3.5	Conventional aesthetical appealing lights
6	SiC	430-505	Blue	3.6	Regular lighting (task and high performances)
7	GaInN	450	White, bluish green. Blue near to UV	4.0	Regular lighting (task and high performances)

Semiconductors are used in making LEDs that can produce different light colors by altering its composition.

It took humans almost some 90 years from switching fire as a light source to the incandescent lamps, then from incandescent to LED – 8 years

A semiconductor that radiates light when subjected to electrical impulses – a phenomenon called electroluminescence.

An acronym for light-emitting diode is LED,

LED runs on direct current (DC) and often requires a separate electrical ballast – a 'driver'. The driver converts the mains voltage to an optimal level for the LED.

Chapter 2

LIGHT & CIRCADIAN

....

It is always motivating to watch people having the enthusiasm in growing their knowledge and our lighting fraternity is advancing in the exponential way in all the aspects, be it technology, flexibility, improvisation on SPD, hue, saturations, efficacy or the efficiencies. Lighting fraternity is also trying to do its best towards the planet earth, however there is a huge gap between learning the significance of the lighting parameters with respect to human physiology.

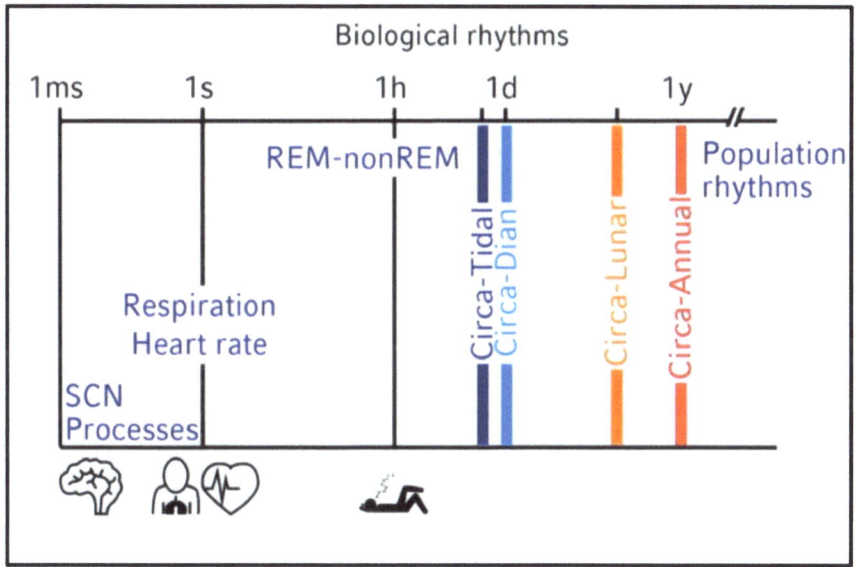

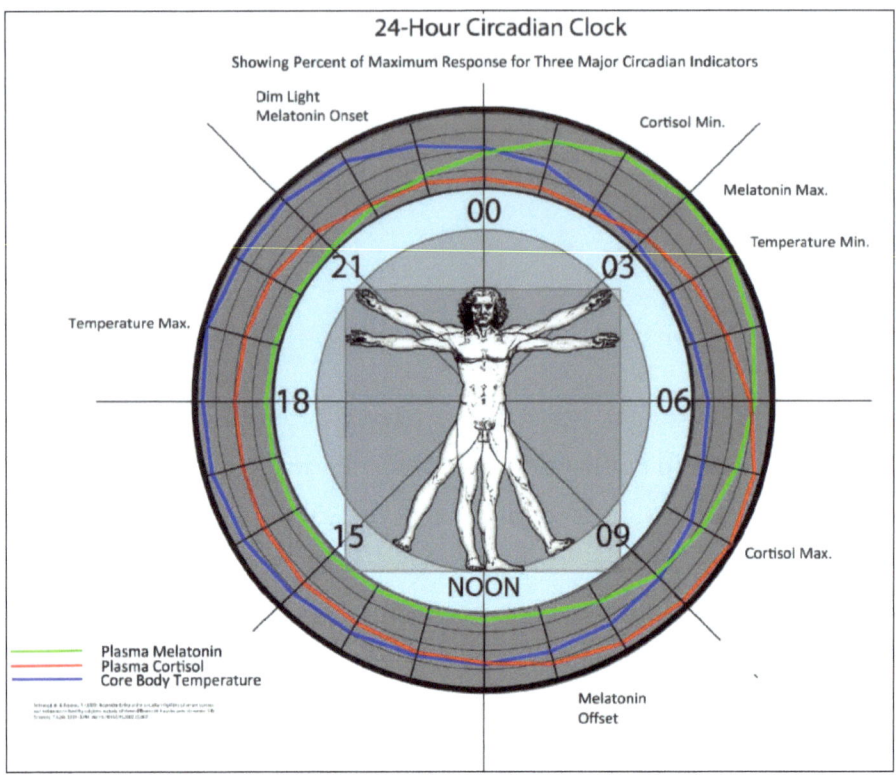

This is because no college/university or the education platforms (pan globe) have included a mandatory at least six months course in the respective curriculum. By introducing this curriculum, not only they will learn, but there will be a very strong bonding of lighting fraternity with the medical industry for the betterment of humans. The same can be applied for horticulture applications.

Hormones scales high & low both with pulsatile patterns but also with longer daily expression patterns. The electrical activity in your brain shows different characteristics over the day and night –this is especially obvious during sleep, but it also has indications on how well we perform over the course of the day.

We are better at solving problems in the very early to early to mid to late afternoon than at other times of day. Do not believe? That's where the we need to be aware that physiology is regulated by the clock. I might be sounding frenzied, it is so pathetic to see that how ignorantly we mistreat our circadian and our sleep needs

I was and am fortunate enough that somehow since past 5 years I got into this learnings, experience & hand on experimentation and finally getting into light therapy which allowed me to study physiology more deeply and closely.

Thus the next chapters will be purely from biology point of view enabling you all to brush up the same. Though you can find the same on any platforms, but I wanted to have this collated with in one bind.

LET'S START

Since light and sleep have very strong connection, we will start with sleep and circadian rhythm (melanopic pattern)

"Sleep is the form of payback towards the fatigue & we certainly have to repay this towards the proper functioning of our body and the brain."

Before we discuss *Circadian Rhythm*, first we need to know the glands, hormones & the neurotransmitters which collectively drives circadian rhythm once the eye/retina is exposed to any form of light (invisible & visible form).

The most important question which one can ask to oneself and to any that how lighting influences the molecular mechanisms, and which part of the molecule mechanism triggers the system generating the effect or impact.

Thus, we need to know the glands, hormones & the neurotransmitters of the human organs associated with this.

GLANDS:

Pituitary gland: The pituitary gland is a small pea-sized gland that plays a major role in regulating vital body functions and general wellbeing. It is referred to as the body's 'master gland' because it controls the activity of most other hormone-secreting glands. This sits inside the brain directly behind or in line of our nose bridge.

Pineal gland: The pineal gland is a small endocrine gland in the brain of most vertebrates. The pineal gland produces melatonin, a serotonin-derived hormone which modulates sleep patterns in both circadian and seasonal cycles. It is located on the back portion of the third cerebral ventricle of the brain, which is a fluid-filled space in the brain. This gland lies in-between the two halves of the brain.

HORMONES:

- **Melatonin – A Hormone of Dusk to Dawn**
1. Defined as, the hormone responsible for sleep.
2. It increases the cells regeneration.
3. Acts as an antioxidant.
4. Stimulates the immune system- making it stronger
5. It is the essential hormone for chronobiological adaptation.
6. The SCN-activated thru Pineal gland, light-inhibited production of melatonin conveys the message of darkness to the clock and induces night-state physiological functions like sleep/wake
8. Blood pressure and metabolism.

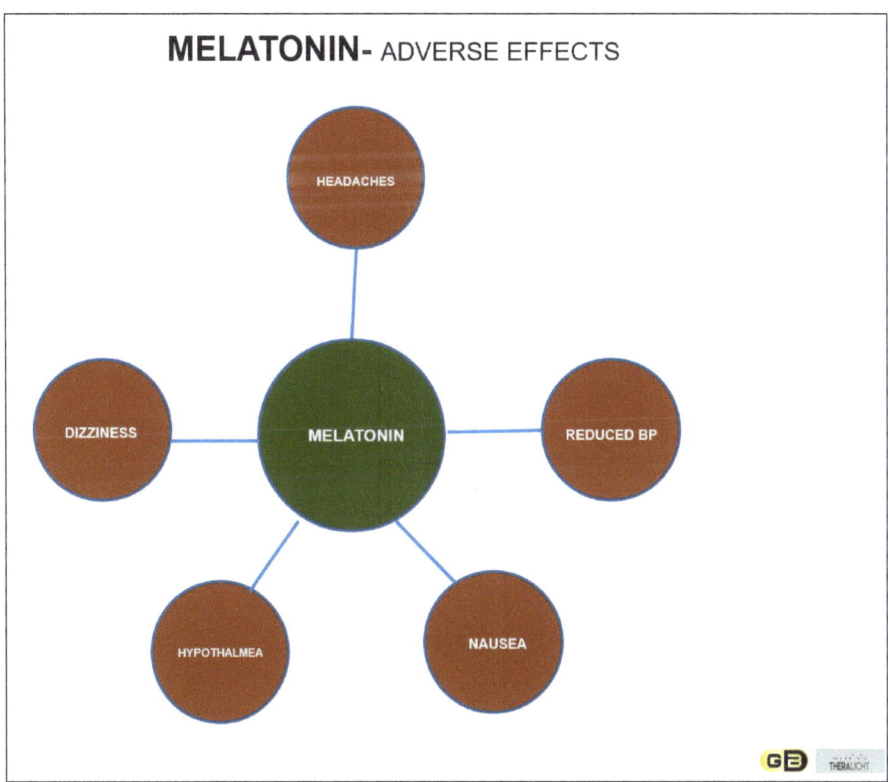

- **Cortisol – A Steroid Hormone**
1. Defined as, the Steroid hormone responsible for the alertness
2. It triggers thru the Pituitary gland to take control of our alertness activities. It is released with the pattern that recurs every 24 hours (Diurnal cycle).
3. Boost energy levels
4. Manages Carbohydrates, proteins & fat.
5. Regulates blood pressure.
6. Aligning itself with certain parts of brain like hypothalamus & pituitary gland it controls fear, mood & motivation.

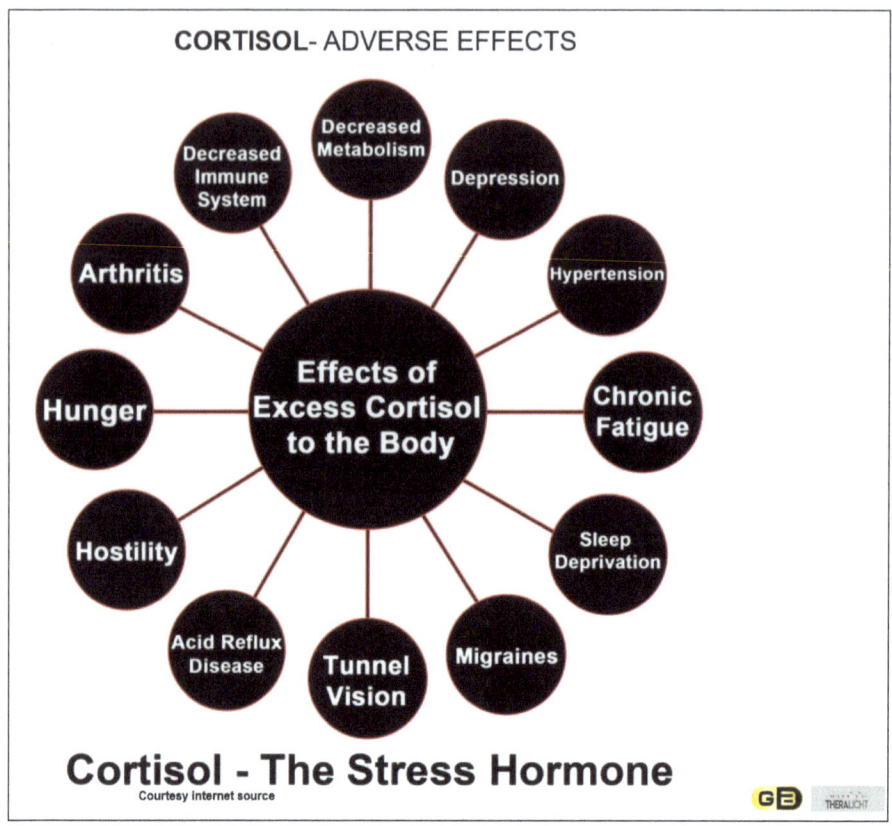

- **Neurotransmitter**

Serotonin is the key hormone that stabilizes our mood, feelings of well-being, and happiness. It also helps with sleeping, eating, and digestion. We have at least 12 types of serotonin. High serotonin levels are associated with high status, rejuvenation and high energy.

Low serotonin levels are associated with bad mood, low energy, low self-esteem and aggression like behaviour.

Dopamine is a chemical found naturally in the human body. It is a neurotransmitter, meaning it sends signals from the body to the brain. The right balance of dopamine is vital for both physical and mental wellbeing.

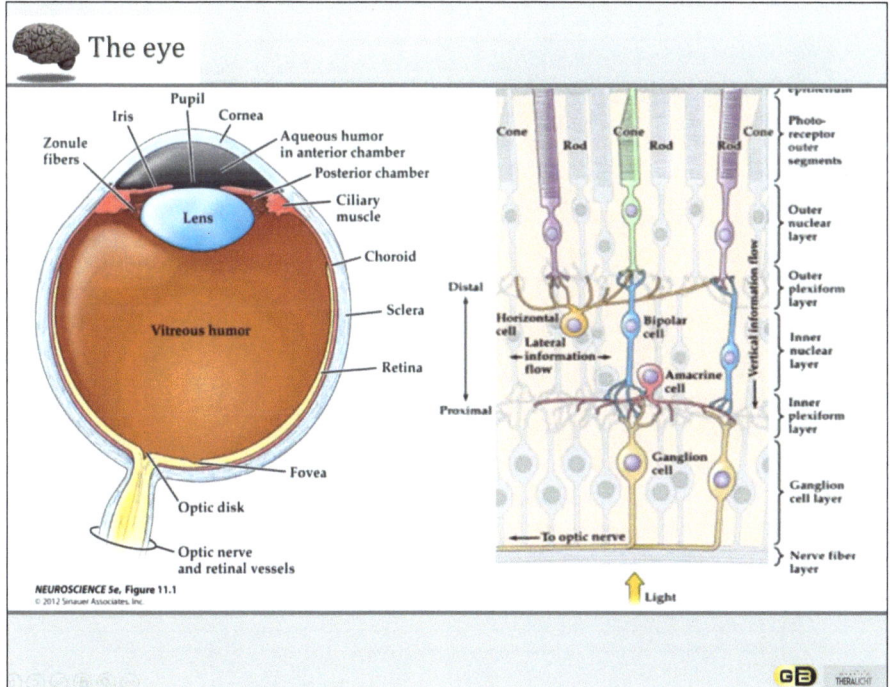

- **Eye**

Let's take dive into the human's brains' brain, THE EYE and its vital parts sensitive to LIGHT.

In the human eye there is a novel photoreceptor system, separate from the rods and cones used for vision, which detects light and is responsible for non-visual responses, such as resetting the internal circadian body clock, suppressing melatonin release and alerting the brain.

These photoreceptors are located in the ganglion cell layer, where there is a specialized subset of cells which are specifically responsive to blue light. Previous research has indicated that these cells are the primary way in which light is detected for the non-visual responses.

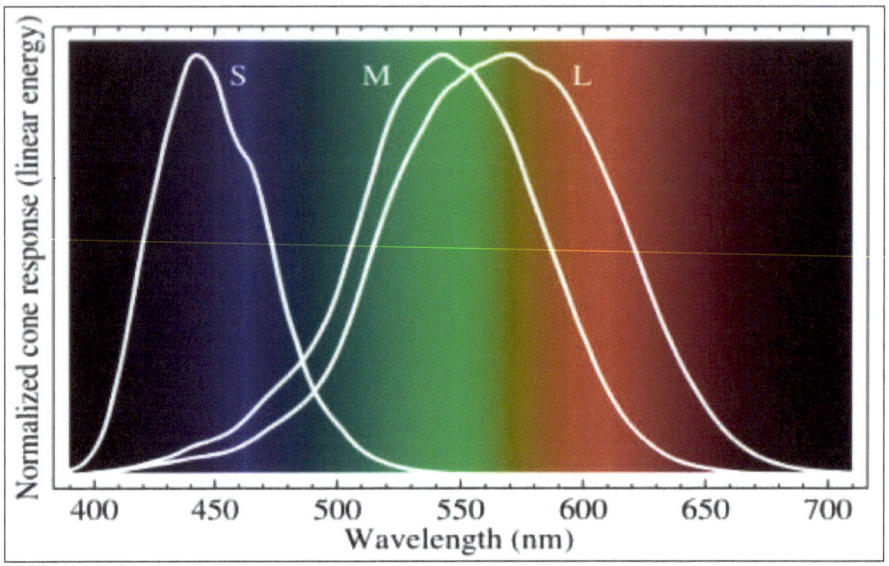

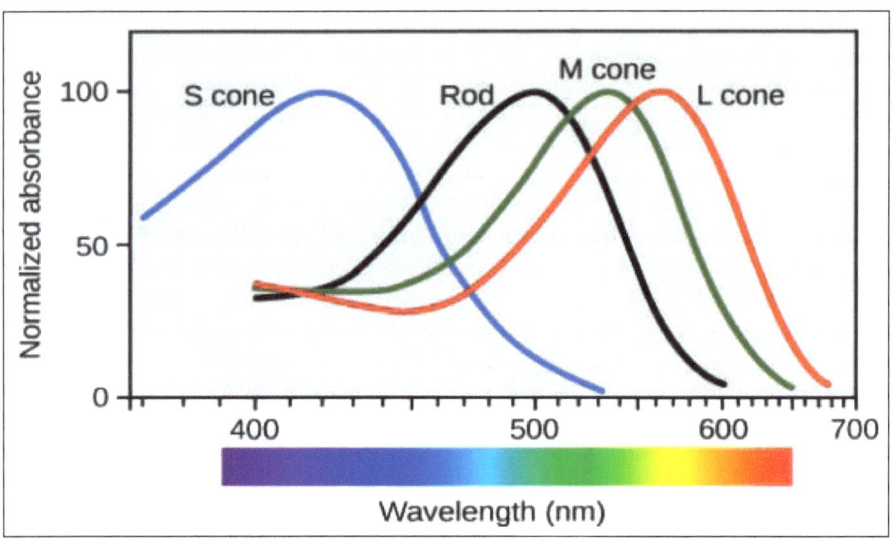

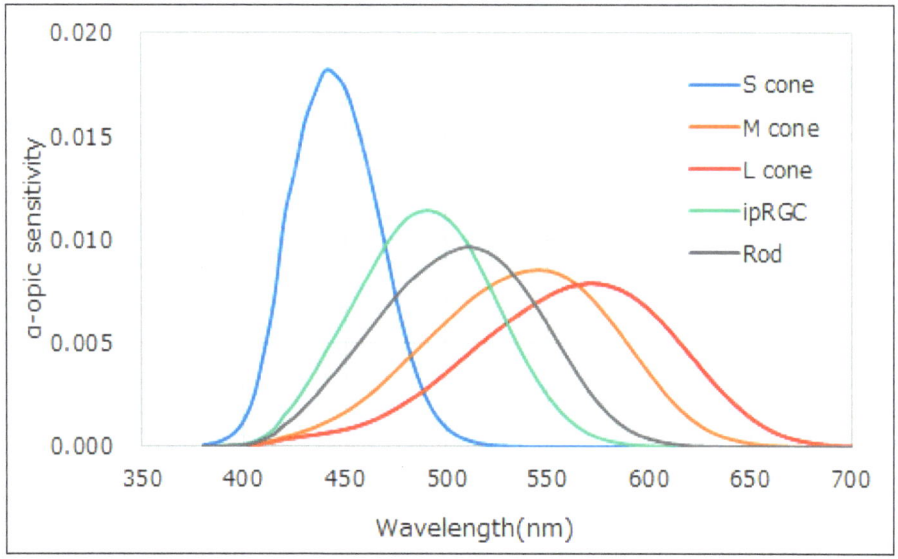

Photoreceptors are specialized neurons found in the retina that convert light into electrical signals that stimulate physiological processes. Signals from the photoreceptors are sent through the optic nerve to the brain for processing.

There are five types of photoreceptors in the human eyes:
- L Cone,
- M Cone,
- S cone,
- Rods and
- ipRGC

Rods are responsible for vision at low light levels (scotopic vision). They do not mediate colour vision, and have a low spatial acuity.

    Cones are active at higher light levels (photopic vision), are capable of colour vision and are responsible for high spatial acuity. The central fovea is populated exclusively by cones.

It is known that the rod cells are more suited to scotopic vision and cone cells to photopic vision and that they differ in their sensitivity to different wavelengths of light.

It has been established that the maximum spectral sensitivity of the human eye under daylight conditions is at a wavelength of 555 nm, while at night the peak shifts to 507 nm.

In visual neuroscience, spectral sensitivity is used to describe the different characteristics of the photo pigments in the rod cells and cone cells in the retina of the eye.

Human's eyes have three types of cones:
- The first L Cone responds the most light of longer wavelengths, peaking at about 560 nm; this type is sometimes designated L for long.
- The second type M Cone responds to the light of medium-wavelength, peaking at 530 nm, and is abbreviated M for medium.
- The third type S Cone responds the most to short-wavelength light, peaking at 420 nm, and is designated S for short. Functionally, only S-cone output can add to the ipRGCs direct response to light. Cells mediating the signals of S-cones are unique in responding to wavelength rather than energy contrast.

The three types have peak wavelengths near 564–580 nm, 534–545 nm, and 420–440 nm, respectively, depending on the individual.

For ipRGC, Melanopsin's are the photopigments used by Retinal Ganglion Cells. The spectral sensitivity differs from that of rods and cones making established photometric measures inappropriate for quantifying effective light exposure for this new photoreceptor.

In addition to bipolar cells, we have horizontal cells receive inputs from the photoreceptors but also gives rise to photo receptors. Thus there is a bidirectional relationship from physiological point of view.

The major part in in inner part of retina is Retinal ganglion cells, these ganglion cells forms the thick horizontal layer across the inner surface of retina. They receive the direct input from the bipolar cells and also receive the indirect input from amacrine cells.

The process of photo transduction requires interaction the photon of light with the photopigments which is a complex molecule, which consists of opsin which is a large protein structure which interacts with smaller organism of molecule called 11cis retinal, it is a member of 7 trans membrane protein family. 11 cis retinal interacts with photon directly.

The solution to this problem in which 'Melanopic' illuminance (m-lux) is estimated by weighting irradiance at each wavelength according to a sensitivity function.

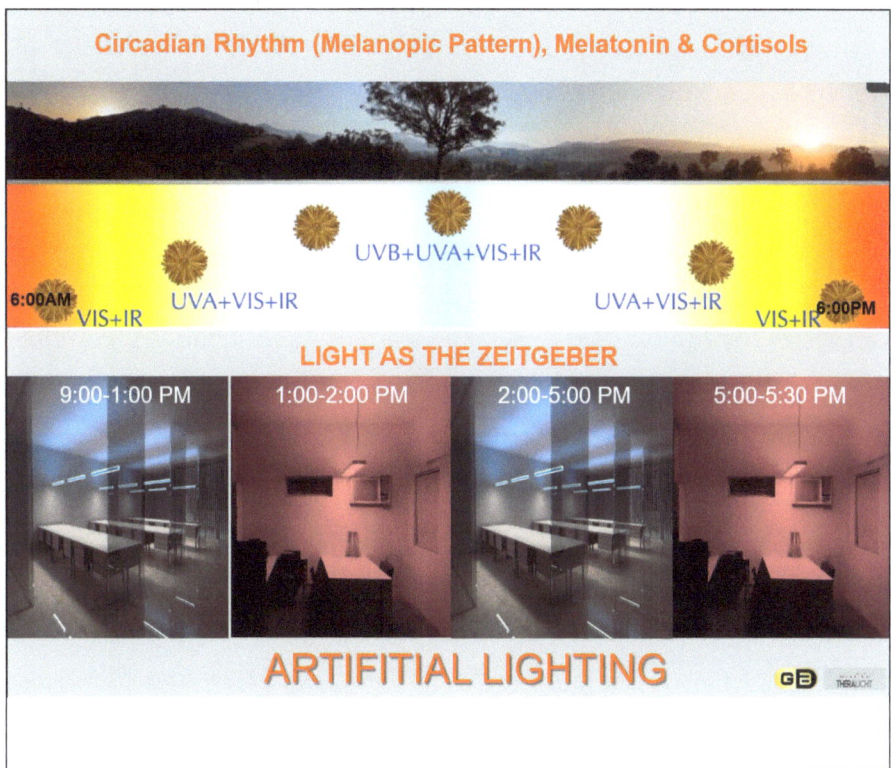

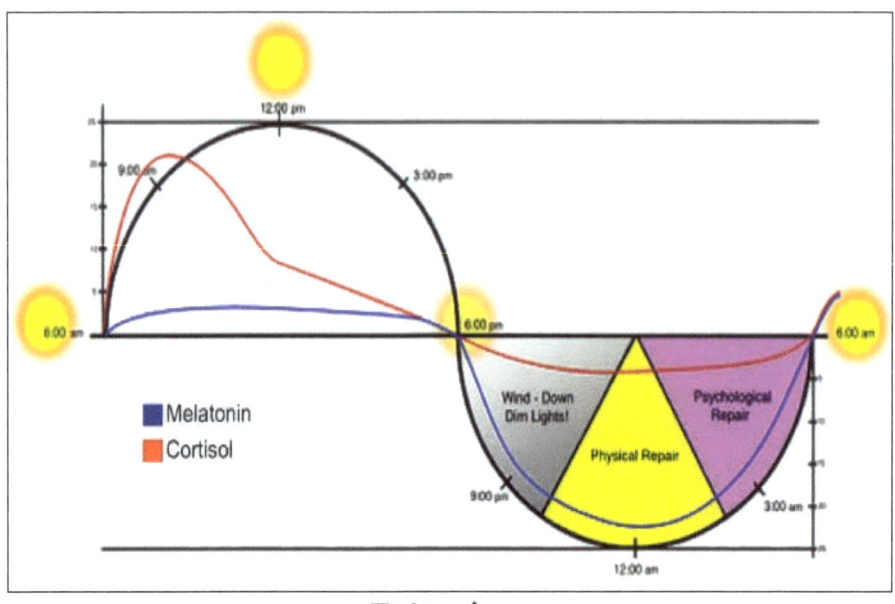

Zeitgeber

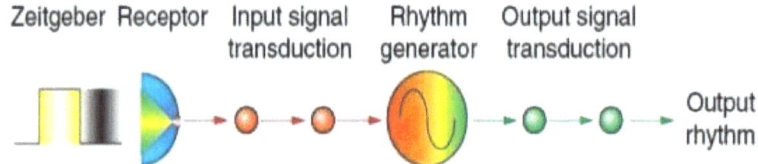

Circadian rhythm or the melanopic pattern in humans: A circadian rhythm, or circadian cycle, is a natural, internal process that regulates the sleep–wake cycle and repeats roughly every 24 hours.

The SCN drives the circadian rhythm in pineal melatonin production (that is, high melatonin levels at night and low melatonin levels during the day) via a multi synaptic pathway that projects to the Periventricular nucleus of the hypothalamus (PVN)

The superior cervical ganglion (SCG). Core body temperature (CBT) and the pineal hormone melatonin are the most commonly used phase markers of this rhythm.

Melatonin is used more often since it is not subject to as many masking influences as it can be measured non-invasively.

This chemical & the electrical process is triggered by the set of the Ganglion cells. Ganglion cells are the final output neurons of the vertebrate retina. Ganglion cells collect information about the visual world from bipolar cells and amacrine cells (retinal interneurons). This information is in the form of chemical messages sensed by receptors on the ganglion cell membrane.

Intrinsically photosensitive retinal ganglion cells [ipRGC] communicates information directly to SUPRACHIASMATIC Nucleus that forms our biological clock mainly responsible for our CIRCADIAN RHYTHM. A central circadian clock, located in the suprachiasmatic nuclei (SCN) of the hypothalamus.

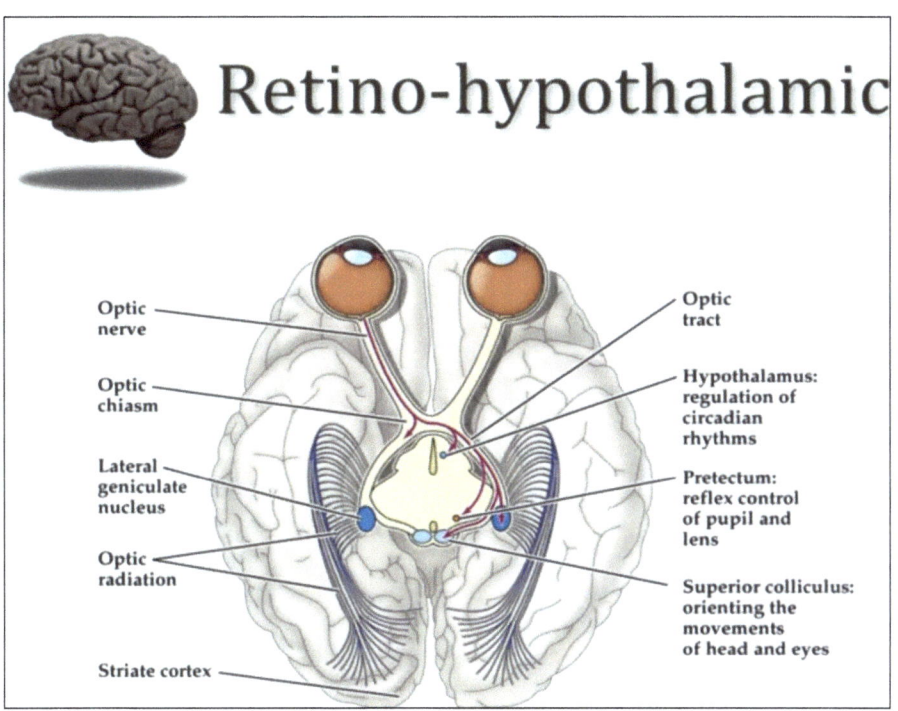

A team led by researchers at the RIKEN Brain Science Institute, Wako, has demonstrated that the neurotransmitter serotonin is the key to this link.

- **A scenario:**

In today's world, the nights are days and the days are nights for many of us. We are exposed to the phenomenon known as ELECTROLUMINESENCE.

Humans receiving high Circadian Stimulus [CS] during the entire workday, exhibited reduced depression and better sleep quality compared to those receiving low CS.

Chapter 3

PHYSIOLOGY & LIGHT

Physiology – Photo bio modulation (Brief): It all begins with photo bio modulations.

Molecules which are chromophores', absorbs the wavelengths emitted by the light source. There is an optical tissue window which varies from 600 nm to 1400 nm (infrared spectrum). The chroma force in tissues which absorbs the light energy. These chroma force are found in MITOCHONDRIA.

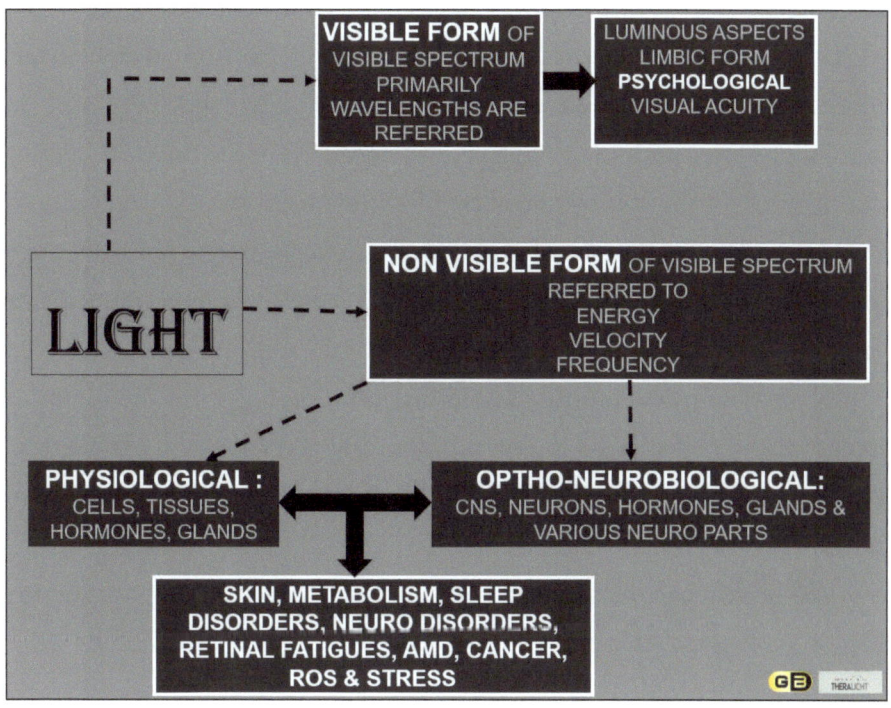

MITOCHONDRIA, in these we have specific molecule known as Cytochrome C oxidase, which is responsible to produce energy within MITOCHONDRIA. This means Adenosine Triphosphate (ATP), which is the end product of energy correlation and is the fuel our cells need for transporting irons for metabolism. If ATP production is stopped, we can die in 15-20 seconds.

Our body produces 30-100 KG of ATP in 24 hours depending upon our weight, height, age etc.

The light is so essential and so much misunderstood or taken for granted that we are ignorant of a very critical and vital fact that the light range from 480-850 nm can boost the energy in the cells where the energy is getting depleted.

The mechanism can devised where the long part of light wavelength can be used to boost energy levels in MITOCHONDRIA. The other hidden fact is that only 1/3rd part of the energy in our body comes from the food rest 2/3rd comes from the radiations and the light.

LEDs don't have the fed frequency with 500-900 nm that hits the Cytochrome C oxidase. We must differentiate the metabolically used or consumed energy with the energy being received from the thermodynamic aspects, which in turn is regulated by EMW w.r.t. EMFs.

Our body's temperature of 37 degree Celsius (98.6 degree F) is not only the result from MITOCHONDRIA but also from the longer wavelengths (nearly infrared) and the pho-energy wavelengths too.

In physiology, we must be aware that our limbic part of the brains is psychologically effected by the types of lighting and its related parameters, like, Spectral power distribution, Correlated color temperature, Gamut index, CRI, the lumen package etc.

The limbic part of the brain can be stimulated thru an appropriate form wavelengths either thru monochromatic light medium or very concise form of SPDs. Generally mentioning, you can relate this to warm colours (2700K/3000K), bright white (6500K-8000K) colours and daylight (4000-5500K) colours when it is referred for the mood lighting.

The arrangements of light also plays a crucial role to stimulate hypothalamic part and thus the limbic one.

The limbic part of the brain is constituted of
1. Amygdala: The amygdala is a collection of cells near the base of the brain. There are two, one in each hemisphere or side of the brain. This is where emotions are given meaning, remembered, and attached to associations and responses to them (emotional memories). The amygdala is considered to be part of the brain's limbic system
2. Hippocampus: Hippocampus is a complex brain structure embedded deep into temporal lobe. It has a major role in learning and memory. It is a plastic and vulnerable structure that gets damaged by a variety of stimuli.
3. Cingulate cortex: The cingulate cortex is usually considered part of the limbic lobe. It receives inputs from the thalamus. It is an integral part of the limbic system, which is involved with emotion formation and processing, learning, and memory.

Apart from the afore-mentioned physiological parameters which are light dependant one of the receptors from 7 TM receptors is also light dependant. 7-Transmembrane receptors (7-TM receptors), also known as G protein-coupled receptors, or GPCRs, are integral membrane proteins that contain seven membrane-spanning helices. Upon ligand binding, the GPCR undergoes a conformational change which is transmitted to the G protein causing activation. Further signal transduction depends on the type of G protein. One of those are activated by a wide variety of ligands including light, olfactory stimulants, peptides, hormones and neurotransmitters. They are grouped into 6 classes based on sequence homology and functional similarity: one of them is Class A (or 1) (Rhodopsin-like)

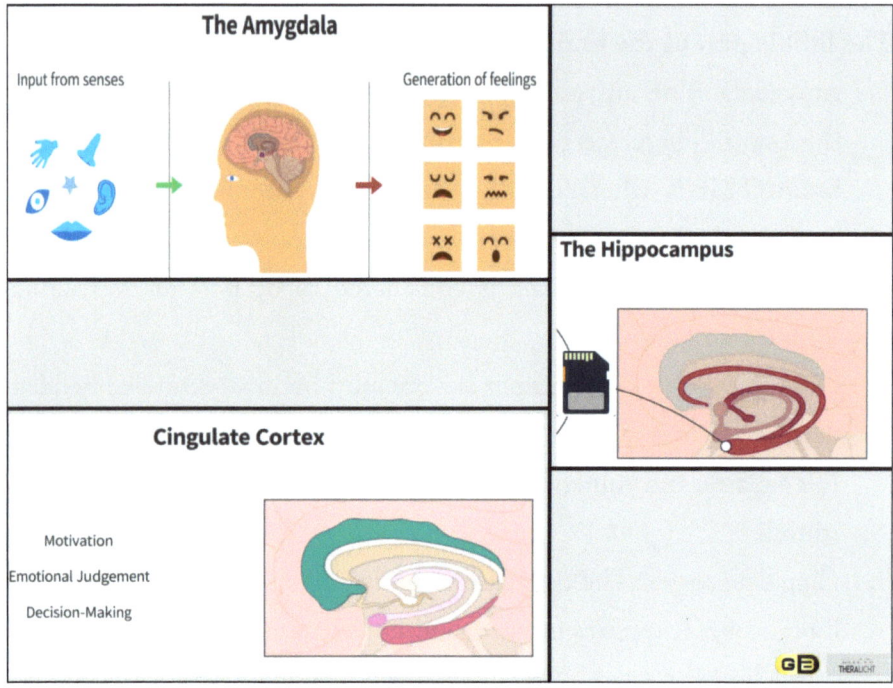

I will like to refrain myself of advising how to carry out the lighting concepts/ design/consultancy, however will like to put up the suggestion that the picture shown can assist you, or your specifier your clients, or your vendors achieving the correct and desired outcome of the lighting products selected and arranged as per the interiors, architecture or lighting concept.

While selecting or manufacturing the product we must not ignore the frequency and energy exitance parameter of the spectrum (SPD). Though many manufacturers are working around the improvisation of the SPD (like lower *intensity of 480-495nm and an increase in 545-555nm & 600-680nm band width along with the balanced intensity to achieve the desired CCT) from LED source.*

While getting into the design after applying all the probing tools or the respective processes laid by the lighting consultancy firm, one must understand the retinal sensitivity graph w.r.t. the age.

To cite some example(s)

- The challenges for lighting industry are to enhance the cognitive skills, attention and the alertness of the employees while they are in office hours.
- The other and bigger challenge for lighting fraternity is to induce all physiological, psychological and sociological parameters thru an appropriate lighting design and products for the shftwork employees working in IT based companies, as they are forced to work against the natural phenomenon of zeitgebers.
- In residences other than selling or recommending lights based on the aesthetic appearances but also understanding the age and gender related factors.
- Children will need a different set of lighting parameters as compared to the older group of generations.
- Women, who spend most of their time in homes will need a different layout of lighting.
- The office goers returning to their homes will demand skilful layers of lighting.

The huge variation w.r.t. the sensitivity in **the spectral transmission of the human crystalline lens** exists between the ages of 40 and 59 years& the age of 60 and older.

The decrement in transmittance between these two age groups varies from 40% for 420 nm to 18% for 580 nm.

Though age is not the only parameter affecting crystalline transmission. **Moreover, the light transmitted decreases with age.**

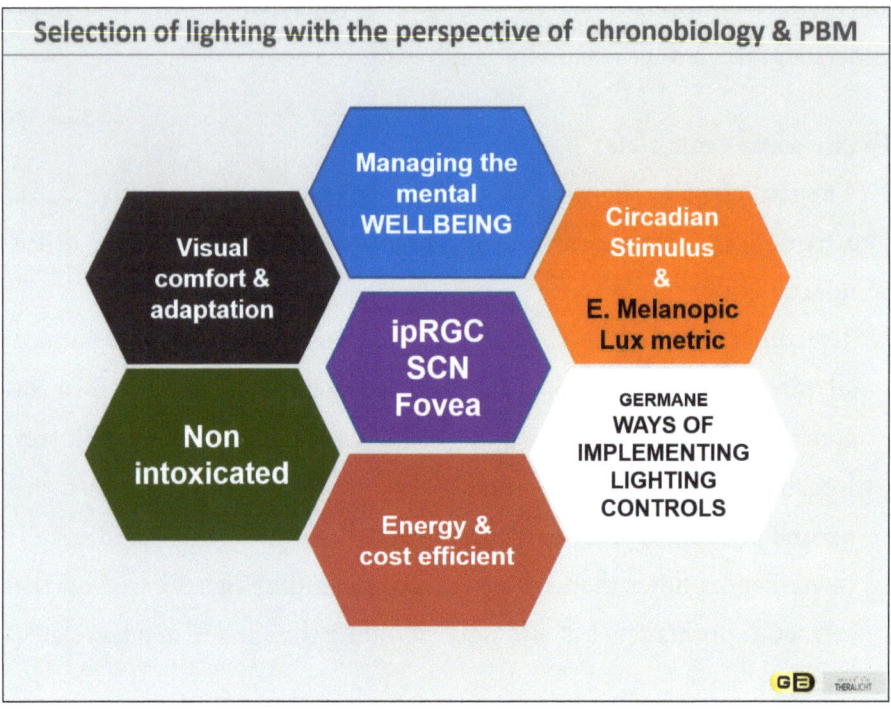

This total transmission of light is similar to or lower than the amount that the different intraocular lenses transmit, even with a yellow or orange filter. The color of the human lens becomes yellowish and saturates with age.

As Antibiotics are essential to treat bacterial infections. LEDs are efficient and provides us the flexibilities in the designs for many applications to be sufficed.

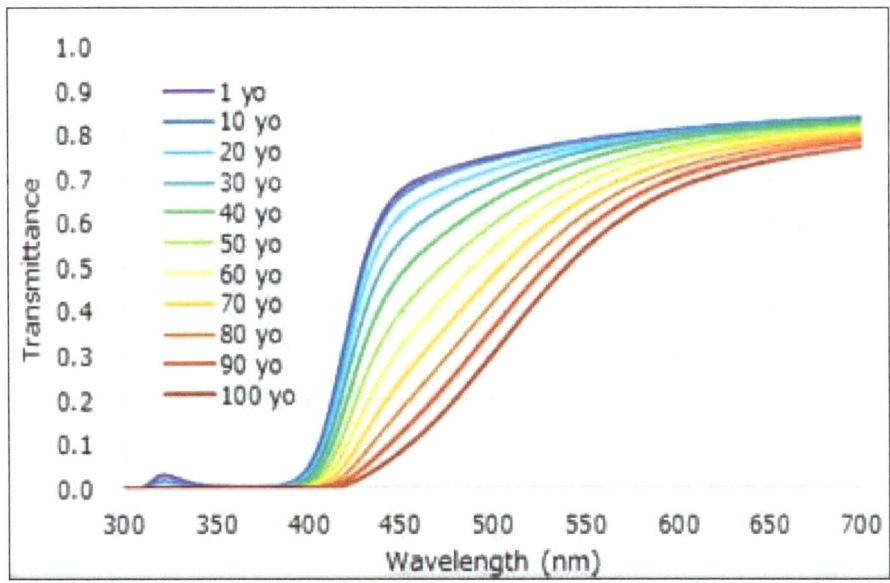

Too much of antibiotics weakens the immune system. The same goes for LEDs- inappropriate usage & farcical lighting designs can be detrimental to the brain, skin and the eyes.

Too much of increase in cortisols is too much of activity & which lowers the immune activities because of which they are unable to kill the cancer cells at an early stage & thus can lead to the cancer.

In the absence of UVB as a good character for VITAMIN-D, now-a-days almost many office goers have the deficiency of the same. Too much of short wavelengths: Retinal damage, brain fatigue and so on.

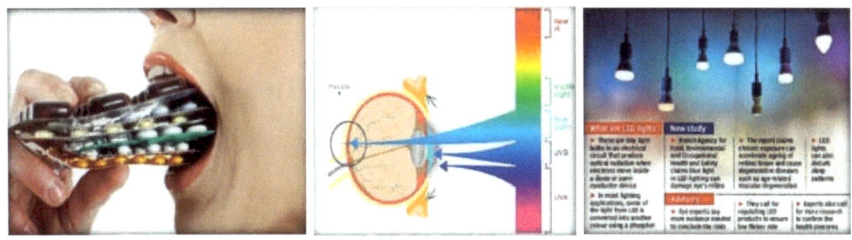

The purpose of choosing an appropriate light and its effect for any building (indoors), green walls or landscaping, is not only the involvement of commercial or technical aspects, but it's the series of physiological & neurobiological factors

Facts being misused

1. Day lighting is a day light from the Sun & cannot be replicated by any kind LED lights. The fact cannot be challenged, overtly, as the day light changes every one hour w.r.t. YOR of the visible spectrum. (Need a proper and accurate system)
2. Implementing tunable white luminaires in isolation is just a placebo. Thus not considering other design & the LED parameters of a product will never have a physiological effects & will not contribute in a major form for CR & CS clinically & biologically. (Please refer to any good physiologist or expert of tPBM). (The customers are now a days are putting the condition what if circadian rhythm is not aligned/generated/triggered?)
3. With the different elevations of the sun and its trajectory, the composition of visible blue spectrum changes which is the prime source of healthy cortisol. Thus the conclusion is, indeed, abuse of artificial light has been associated with harmful health effects in humans such as an increased risk of developing migraine, AMD, insomnia, stress & cancer apart from the psychiatric disorders and obesity. We must use light for our physiological, sociological and psychological benefits and not abuse the same thru our farcical product, design lighting design and bad quality.

Some of the lighting organizations, individuals whether manufacturers or sellers or the consultants claims that the tunable white aligns the circadian processes, this is a myth. The correctness here I will like make is it's just not only the SPD (Spectral power distribution whether R9 values or any other alterations), the introduction of bio red, bio blue and correctness of 585-595 nm are equally important to trigger the Process C of circadian rhythm or the melanopic pattern.

Chapter 4

AN INSIGHT: THE SOLUTIONS

....

As the ear has dual functions for audition and balance, Eyes have a dual role in detecting light for a wide range of behavioral and physiological functions apart from sight (source)

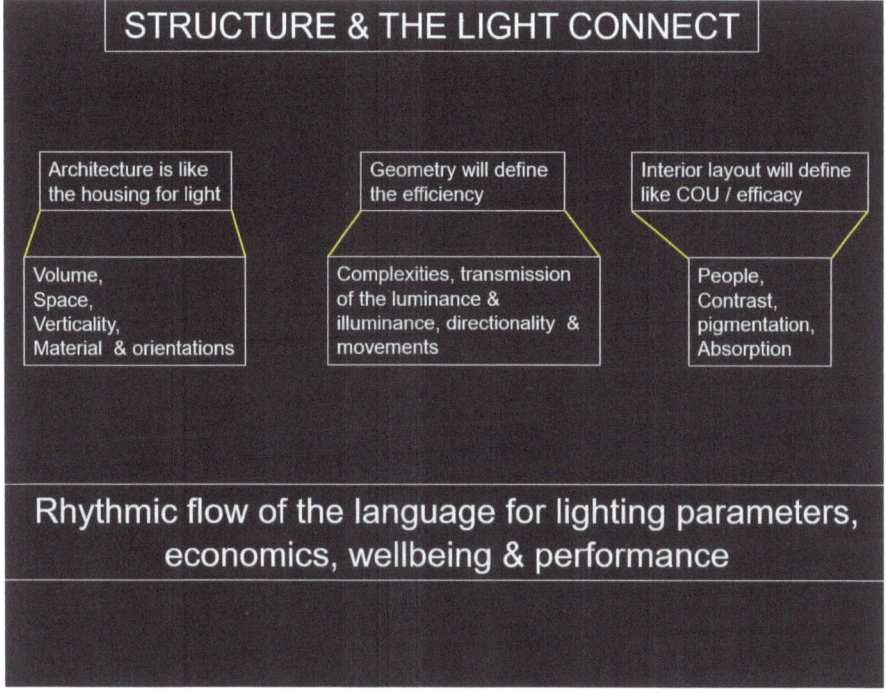

Whether effulgent or subtle. Light needs its own space to illuminate and create shadows…which is an essential part of our daily livings.

The biggest enemy and friend of light is the space around. It is up to us what we choose for our space, an enemy or a friend?

By mentioning this, I need to draw your attention towards our rooms/cabins/open spaces in the homes and offices.

Narrowing on the word "space", primarily is the ceiling itself which can make or break an environment with respect to light and thus affecting our moods, aesthetics or circadian rhythms.

It all started mushrooming in my mind when I was with a pool of the architects and some customers, way back in the year 2009 and we discussed that how the paucity of time, overnight constructions of the buildings & accommodation of large group of employees on yesterday basis has to be dealt?

FALSE CEILING: The impacts of false ceilings reduced volumetric spaces, increasing suffocation and increase in number of lighting fixtures & air conditioners.

Thus, there was a major thrust on direct/indirect lighting which really improved the overall parameters of the environment w.r.t the productivity, liveliness and total cost of ownerships by reducing the quantum of the luminaires used as compared to the earlier conventional methods of down lighters or the recessed lighting

IGNORANCE IS OUR ENEMY:

The lighting industry on its innovative path was trying to define another horizon for the benefits of human being but THE ENEMY- FALSE CEILING continued to be the preferred one with more options like Metal, mesh, Gypsum, Armstrong and lately the Techzone ceiling.

However the design-knack is one of the option to deal with this and many manufacturers have done the excellent job of designing the luminaires to compensate the negative effect from these recessed luminaires.

No one from the lighting industry (as a vendor or manufacturer) has an audacity to question the specifiers or the customers that for what best reasons other than aesthetics, accommodating the air conditioners, its cables, pipes and lighting fixtures (primarily), false ceiling is being used? (Exceptions are always there)

Even today, it is like a herd of sheep, who don't even know, why they are following one another and where?

I started observing, discussing and speaking to the respective industry enthusiasts & to my surprise they were (and are) on the same page, that usage of FALSE CEILING has no meaning and there can be many other ways to compensate the aesthetics.

Why LIGHT NEEDS SPACE? Light needs space to travel with a smoothest way in the room (especially bedrooms, offices, hospitals rooms & hotel rooms) to provide the solace to the entire atmosphere.

When I entered the ring of challenges with all the enthusiasms to elicit the solution, I was helped and assisted by my friends from the lighting fraternity. (Tashi Aggarwal, Vishal Kapadia, Sushant Surve, Amanpreet, Vinit Narula, Shashank & Abhishek Khandelwal)

My research work and product in line to align circadian rhythm, the physiological parameters like immune system & ATP production will help office people whether in offices or from their respective homes. The product has compensated the missing parameters of LED with the conventional way of manufacturing and arranging the LEDs. I have given it the name SALVATION series.

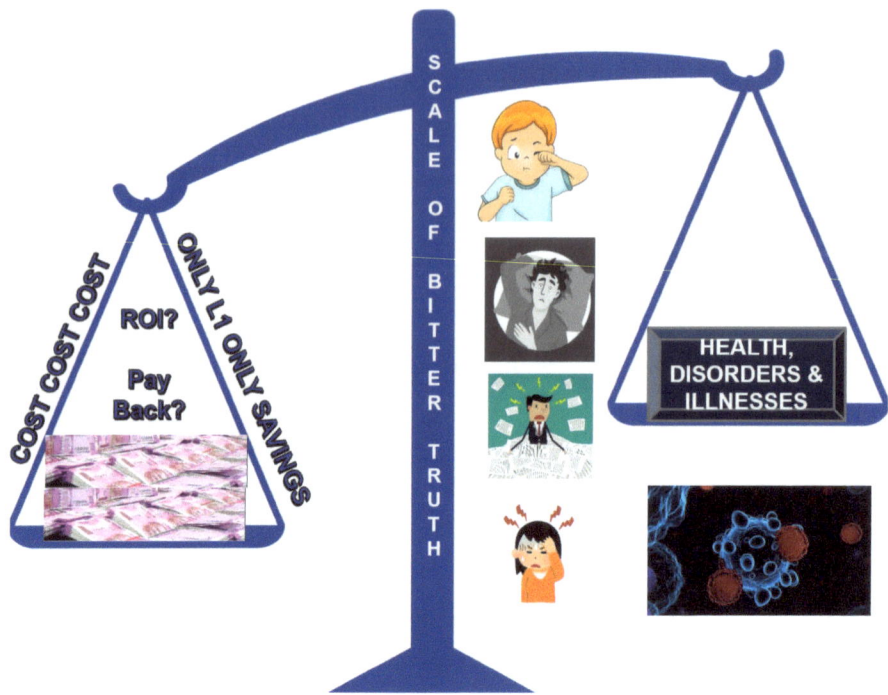

- **Study Objectives:**

To examine the association between bio-red, bio blue, NzER, FzIR and swift switchovers of Kelvins (2700-5500K) during office hours.

- **Methods:**

A total of 19 office goers who were subjected to SALVATION SERIES. From morning 9:30-10:00 AM were subjected with blubio+4000K trigger the pituitary gland

From 10:00 AM-1:30 PM they were subjected under bio-blue+5500K

From 2:15 PM-2:30 PM they were subjected under bio-red (NzIR+680).

From 2:30 PM - 4:30 PM they were subjected under bio blue+5500K

From 4:30PM—5:30 PM they were subjected under 3000K

In the present study as baseline and were followed up to 25 to 45 years of age. Sleep duration was categorized as ≤ 10 hours,

- **Results:**
   The evident incremental in cognitive performance, reduced stress, better sleep nap & over all acknowledgement of impact of time cues

- **Introduction:**
   Because all light sources were adjusted to provide the same horizontal as well as vertical illumination from table till 25 inches on a vertical plane (the standard measure that lighting designers use to illuminate indoor spaces), and because the spectral power distributions of the chosen LED light sources were significantly diverse, the intensity of total visible

   Light entering the eye (measured vertically at eye level with the subjects seated at the conference table) varied between the light sources. This provided the opportunity to determine if the melatonin which was influenced by corneal light metrics (irradiance in μW/cm2, photon irradiance in photons/cm2/s, photopic illuminance in lux). Individual pupil sizes and macular pigment optical density were not measured and were not taken into account. There was a significant correlation between light intensity by any of these metrics and the melatonin results.

   We next examined how well the most commonly used circadian lighting calculators, the circadian stimulus (CS; Rea and Figueiro, 2018), CIE melanopic irradiance (CIE-026S), and the closely related equivalent melanopic lux (EML) calculator the occupants were subjected under these conditions of variable LED light sources.

- **Conclusion:**
   Though our spectral sensitivity is at 555nm.Sixty percent of the total Circadian Potency spectrally sensitive irradiant power fell within the 438- to 493-nm FWHM band.

   To characterize the transient behaviour of circadian spectral sensitivity, we compared the steady-state Circadian Potency spectral sensitivity curve (with 12-h light exposures) with the best-fit curves for the published experimentally derived data for 15-min bio-red light exposures and for 60-min light exposures bio-blue+5500K.

Morning hours were 210 minutes light exposures with bio-blue+4000K, 4-parameter optimized curve-fitting methodology of our studies.

LIGHTING & HEALTH EXPERTS SELECT SALVATION SERIES

D P Analytics office testifies on the basis of the results and impacts that "This is the office luminaire we've been waiting for"

When **D P Analytics** moved its Mumbai office to a new location they enlisted healthy lighting Design, their in-house group of illuminating experts, to solve the projects lighting challenges. The new space in the Ahmedabad's based Office Tower is a conventional mix of open-area workstations, private offices and conference rooms with low 11' 5" ceilings throughout.

The project's lighting objectives were established "to provide wellness, healthy, interactive, an ergonomic and visually luminance balanced workspace," means, "That evenly illuminating the ceilings, walls and work surfaces, while CS & EML levels the space and lighting should help the occupants with the better cognitive functions, minimizing glare and unwanted reflections."

Considering the limitations of traditional recessed fixtures, these objectives were easier said than done. The challenge was also to create aesthetically bright environments

- **Case Study**

Until now we were concerned with defined lux levels, tuneable white light, UGR, CRI and the w/soft. Now there's a product that offers controlled CS. CR ratio, EML, retino-hypothalamic activation to trigger pineal &

pituitary glands with the designed SPD for special bio-red & bio-blue technology in line with irradiance value to help out physiology of the body, cognitive functions brightness, uniform distribution and exceptional energy performance.

Project: *DPA*
Challenge: *How to improve cognitive skills, align circadian and have a rejuvenated atmosphere?*
Solution: *Salvation series™ recessed luminaires with Patent technology; for CR, CS & EML.*
Results: *The purpose of choosing an appropriate light and its effect for any building (indoors), is not only the involvement of commercial or technical aspects, but it's the series of physiological & neurobiological factors*

Health & Lighting Experts Select Salvation series

The project's lighting design was further complicated by its ambitions for energy performance. DPA was seeking a physiologists & psychologists feedback with a voluntary system for wellness & health.

DPA was assisted by Theralicht™ LLP to carry out the audit for the same with some of the toughest assessing parameters, like DLMO, PstLM, and adherence to CS026 With demands for quality visual ergonomics running up against restrictive energy codes,

DPA's offices were precisely the kind of space that test the limits of conventional lighting. "If we would have went with traditional recessed lighting, I knew we would have to compromise somewhere."

The research on wellness, health and physiology led DPA's team to consider Salvation series™, a new family of recessed luminaries from

My colleagues didn't believe that a light can be a medicine to keep us away from many of visual and non-visual Impacts on our health and body.

"People walk into our offices now and feel that they're in the most appropriate, productive, stress free (limbic) & natural environment."

Salvation series was the best investment we could have made. This is the office Luminaire we've been waiting for."

…Salvation series is the best investment we could have made.

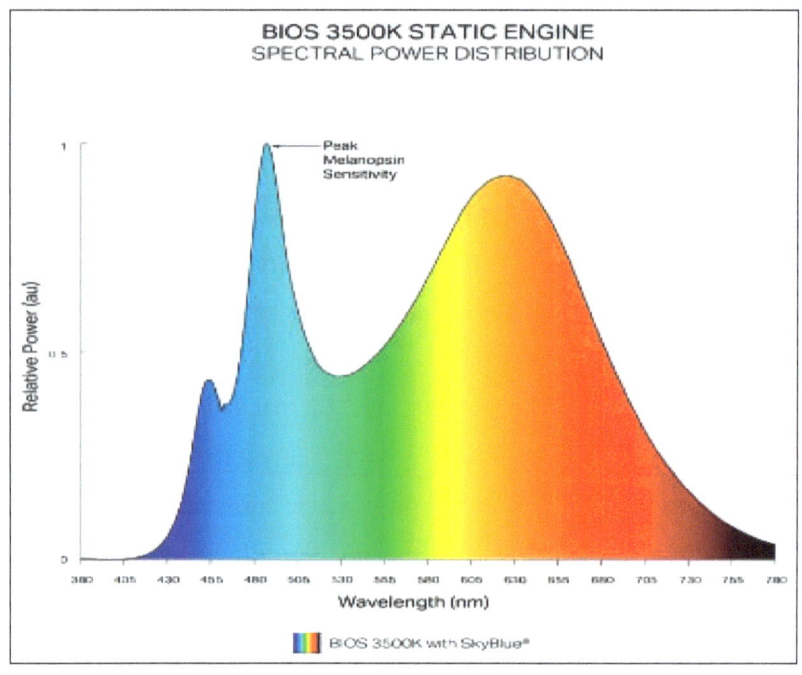

I will like to conclude this chapter of mine with the suggestion:

> "The purpose of choosing an appropriate light and its effect for any building (indoors), green walls or landscaping, is not only the involvement of commercial or technical aspects, but it's the series of physiological & neurobiological factors"

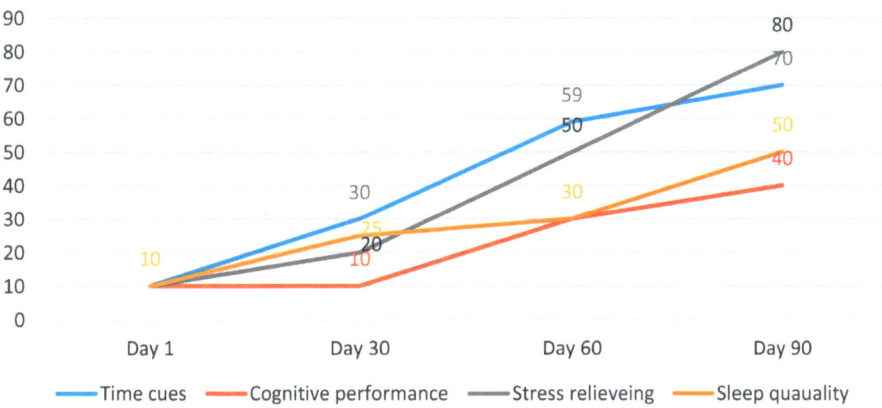

Analysis of the physiological & psychological parameters : Salvation series

Chapter 5

LIGHT THERAPY

. . . .

Over couple of years, with all the challenges of understanding, experimentations, getting the appropriated devices, finances and the paucity of the time (since we are also involved in our respective jobs and businesses as all others), we (Medacsis) took this challenge and waded with all the conviction and enthusiasm.

We formed the team in the year 2019 (Tashi Aggarwal, Ashish Sethia, Vishal Kapadia, Sandeep Behl, and Dr. Vivek Sharma) as we were deeply convinced with the phenomenon and criterion of the wavelengths (680-940nm) which are being used to treat the respective illnesses and the disorders.

Our conviction and the establishment of the light as one of the most prominent source for healing took over the obscurity of all the skepticisms.

Making you all aware of (Medacsis's Low Intensity longer wavelengths) LILW's potential to induce cellular effects through accelerated ATP production and the mitigation of oxidative stress. In clinical use, however, it is often difficult to predict patient response to LILW. It appears that cellular reduction=oxidation (redox) state may play a central role in determining sensitivity to LILW and may help explain variability in patient responsiveness.

In LILW, conditions associated with elevated reactive oxygen species (ROS) production, e.g. diabetic hyperglycemia, demonstrate increased sensitivity to LILW. Consequently, assessment of tissue redox conditions in vivo may prove helpful in identifying responsive tissues.

A noninvasive redox measure may be useful in advancing investigation in LILW and may one day be helpful in better identifying responsive patients. The detection of bio photons, the production of which is associated with cellular redox state and the generation of ROS, represents just such an opportunity.

In this review, we will present the case for pursuing further investigation into the potential clinical partnership between bio photon detection and LILW.

Brain and LILW:
- ✓ The brain receives light. The nature and intensity of this light conveys information which is compared with established memories. The outcome of this process determines or influences our subsequent behaviour, speed of motion, etc.
- ✓ The unique energetics of light activates specific biochemical processes throughout the body – which are essential to maintain the body's physiological stability.
- ✓ Most proteins absorb and emit light. The spectrum of light absorbed and emitted varies according to the nature of the prevailing biochemistries and pathologies. This serves as a mechanism to determine the onset of pathologies from
- ✓ The brain is also a biochemical entity. Its function is influenced by any biochemical influence i.e. by genotype or phenotype. This means that the function of the visceral organs influences brain function and vice versa. It also means that visceral influences alter our behaviour.
- ✓ The function of the brain is not clearly understood. The current understanding of 'executive functions' which are regulated by the

prefrontal cortex is not able to clearly define the nature and significance of these 'executive functions'. It may make more sense to redefine the role of the prefrontal cortex and the executive functions to be instead the regulation of the physiological systems and hence of the body's physiological stability. This appears to be entirely consistent with observed phenomenae and prevailing research.

- ✓ An understanding of the nature and structure of physiological systems serves as a means to predict the precise organs which could be influenced by systemic instability.
- ✓ The structure of the body's function is not yet clearly understood. It is recognized that the organs work in clearly defined organ networks however this is a hugely under-researched area. The nature and significance of these systems is now established.
- ✓ Memories are stored as unique electrochemical signals within the neurons, axons and synapses. The brain stores its memory of internal and external events. It uses EEG frequencies to store and access memories of different physiological significance, and to regulate the stability of the physiological systems. If physiological stability cannot be maintained e.g. due to exposure to environmental stressors, it looks for the best-fit solution. This influences the function of the somato-sensory organs i.e the ears (hearing), nose (smell), mouth (taste), skin (touch), vocal chords (speech) and eyes (vision).
- ✓ There is a flow of information to and from the brain/between the brain and sense/visceral organs. This explains how in rare cases a patient has been diagnosed with little brain and yet they function normally.
- ✓ Finally, our unique genetic profile pre-determines the balance of psych emotional factors in our personality. The emergence of pathologies influence our psych emotional balance and ultimately our behaviour.

Though we also found out that while blue-light is usually the most effective way to stimulate the non-visual responses - especially under bright light conditions- stimulation with green light was also capable of eliciting the non-visual responses under certain circumstances.

However focusing on the benefits of LILW – Medacsis funded research work: "Our conclusions suggest that by exposing NIR. R (620 - 840nm), with the exact duration and pattern of NIR/R light, LILW light therapies could be optimized to cure migraine, insomnia, stress and tensions"

It is also found out that it can be effectively used as the supportive treatment for PD, TBI (some conditions), PSP & EFD

These findings have the potential to play an important treatment role for a number of disorders including circadian rhythm sleep disorders, seasonal affective disorder and dementia.

This study was funded by the by the owners of Medacsis partners

HOW DOES THE IMMUNE SYSTEM WORK?

Our immune system's prime function is to keep us healthy by preventing from the infection. This is accomplished by identifying and destroying harmful pathogens, including viruses, bacteria, and more.

We can accomplish that by using the color red. It's called ultra-red light therapy. It uses light to treat inflammation, wounds, aches and pains, and strengthen your immune system.

It goes unmentioned that the body comprising of 7×10^{27} atoms (approx) must be healthy on a cellular levels. Immune system can be boosted by many ways naturally available with us

The human energy system is in a continual ever changing & in holistic state. This harmony is accomplished by constantly synchronizing our body's atoms, glands, genes, molecules, cells hormones in alignment of time cues- the LIGHT.

For example, UVA and UVB light are both highly therapeutic for the skin and body in general, though the major downside is that too much exposure can cause accelerated aging and even cancer. Other alternatives, such as green light, are not harmful, but also don't provide any substantial benefit.

POWER OF RED, IR, NIR & FIR FORMS OF LIGHT

Medacsis-TheraLicht LLP™ LILW or LLLT, or PBM— with specific wavelengths of light between 630 and 700 nanometers.

Improvement in liver regeneration. Red light stimulates liver regeneration, enabling the liver to "sort" nutrients and toxins and determine which should be absorbed or eliminated.

Stimulation of the lymphatic system, which is critical for a strong immune system.

Activation of stem cells which are then mobilized to become active in the immune response.

NIR/FIR accounts for approximately 50% of the solar radiation reaching the ground at sea level. It has been divided into three bands: IR-A (760–1400 nm), IR-B (1400–3000 nm), and IR-C (3000 nm–1 mm). IR radiation can penetrate the epidermis, dermis, and subcutaneous tissue to differing extents depending on the exact wavelength range being studied.

Exposure to IR is perceived as heat. The strength of electromagnetic radiation depends on the energy of the individual particles or waves as well as the number of particles or waves present.

- A Case study

  Influence of Low intensity light waves on neuronal based activities in TBI patient

  www.medacsis.in www.theralicht.com

  Author: Girish Bhardwaj – Theralicht™ LLP

- **Introductions:**

The case of a baby girl aged 7 years (2019) with TBI with war off time of (2017) 2 years was to be dealt. With the reworked history the baby girl was suffering with TBI due to high fever causing

(i) Severe cerebral and cerebellar atrophy, including atrophy of thalami. Apart from this the white matter changes in the bilateral periventricular white matter of questionable significance

(ii) Large areas of restricted diffusions in almost entire cerebral parenchyma. Bilaterally.

(iii) Mild leptomeningeal, with sever meningoencephalitis

She had lost the movements of limbs, speech, controls, recognition and cognitive functions completely with severe neuronal damage. The parents and the respective doctors had done whatever the best in their capacities and abilities.

- **Aim:**

Our main aim was to trigger and normalise cerebral blood flow along with the regeneration of affected thalamic and respective regions. The aim was also to observe the retinal movements while reducing the Oculomotor Dysfunction which got developed coz of TBI.

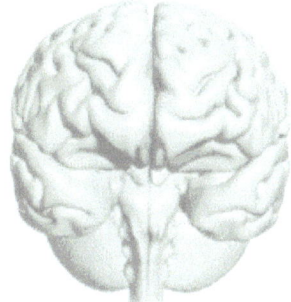 

- **Method:**

The non-invasive low intensity light therapy was considered for the next couple of months. The default mode network-DMN plays critical central role in normal brain activities, presenting greater relative deactivation during more cognitively demanding tasks. After deactivation, it allows a distinct network to activate. This network (the central executive network) acts mainly during tasks involving executive functions.

The procedure of red or near-infrared (NIR) light to stimulate or regenerate tissue is known as photobiomodulation. It was discovered that NIR (wavelength 710–940 nm) and red (wavelength 600 nm) (LEDs) are able to penetrate through scalp and skull and have the potential to improve the subnormal, cellular activity of compromised brain tissue.

Based on this, different experimental and clinical studies were done to test LED therapy for TBI, The patient was subjected to LILW

(690-910 nm) on alternate day basis with head wrap around device for 20 minutes in the morning along with the spine cord subjected to (640-840 nm) on alternate day basis. Along with this the patient was given thyme-leaved gratiola (Btahmi) and promising results were found. It leads us to consider developing different approaches to maximize the positive effects of this therapy and improve the quality of life of TBI patients.

- **Results:**

The significant improvement of holding saliva, reduction in Oculomotor Dysfunction was noticeable in 6 months. The therapy was continued and in a span of 14 months from the commencement of the therapy (still ON) the movement of the limbs, improvement in metabolism, speech, cognitive functions and recognition have been observed.

The Medacsis-LILW therapy is being continued with the hope that this baby girl will be back to some normalised condition in next 1-2 years from now (March 2021

- **Support:**
Conclusions:

It is essential to have a rehabilitation plan, considering the morbidity due to TBI, preferably multi-professional, in order to get the maximum degree of recovery of neuropsychological activities. tLED therapy is an interesting approach to this, since it allows researchers to study the activity of the human brain in real time and finally to guide the patient's brain to plastic changes.64

There are different approaches for treatment of neuropsychological symptoms after TBI beyond PBM. Pharmacological treatments target the modulation of major neurotransmitter systems – dopaminergic, serotonergic, noradrenergic, and acetyl-cholinergic and glutamatergic. Serotonin reuptake inhibitors act for depression secondary to TBI. Modulation of the dopaminergic system improves alertness, attention and cognitive processing speed. Cognitive and memory impairments may reflect the disruption of cholinergic function, and the effects of anticholinergic agents support this contention.65

- **References**
1. Corrigan JD, Selassie AW, Orman JA. The epidemiology of traumatic brain injury. J Head Trauma Rehabil. 2010;25(2):72–80. [PubMed] [Google Scholar]
2. Selassie AW, Zaloshnja E, Langlois JA, Miller T, Jones P, Steiner C. Incidence of long-term disability following traumatic brain injury hospitalization in the United States. J Head Trauma Rehabil. 2008;23(2):123–131. [PubMed] [Google Scholar]
3. João Gustavo Rocha Peixoto dos Santos, Wellingson Silva Paiva, and Manoel Jacobsen Teixeira Alway Y, McKay A, Ponsford J, Schönberger M. Expressed emotion and its relationship to anxiety and depression after traumatic brain injury. Neuropsychol Rehabil. 2012;22(3):374–390. [PubMed] [Google Scholar
4. Schönberger M, Ponsford J, Gould KR, Johnston L. The temporal relationship between depression, anxiety, and functional status after traumatic brain injury: a cross-lagged analysis. J Int Neuropsychol Soc. 2011;17(5):781–787. [PubMed] [Google Scholar]
5. Donnell AJ, Kim MS, Silva MA, Vanderploeg RD. Incidence of postconcussion symptoms in psychiatric diagnostic groups, mild traumatic brain injury, and comorbid conditions. Clin Neuropsychol. 2012;26(7):1092–1101. [PubMed] [Google Scholar]

The Medacsis team since the year 2019 is reaching to the patients who are convinced for non-invasive LILW therapy to heal Insomnia, Migraine, PSP (Progressive Supranuclear Palsy), Parkinson's disorder, stress, muscular atrophy, pains & Traumatic brain injury (TBI*)

I am fortunate enough of getting the motivation and the support from the patients and their kins., thus we are able to further our studies and experimentations for the betterment of humankind while encountering the mentioned challenges:

COGNITIVE

Memory, Learning, Reasoning, Speed of mental processing, Judgment, Attention or concentration, Executive functioning problems, Problem-solving, Multitasking, Organization, Decision-making, Beginning or completing tasks.

SENSORY

Ringing in the ears, Recognizing objects, impaired hand-eye coordination, Blind spots or double vision

A bitter taste, a bad smell or difficulty smelling, Skin tingling, pain or itching, Trouble with balance or dizziness, Memory or concentration problems, Mood changes or mood swings, Feeling depressed or anxious.

SOCIAL & EMOTIONAL

Trouble turn taking or topic selection, Changes in tone, pitch or emphasis to express emotions, attitudes or subtle differences in meaning, Difficulty deciphering nonverbal signals, Trouble reading cues

Trouble starting or stopping conversations, Inability to form words (dysarthria), Depression Anxiety, Mood swings

BEHAVIORAL

Difficulty with self-control, Lack of awareness of abilities, Risky behavior, inaccurate self-image Feeling depressed or anxious, Difficulty in social situations, Verbal or physical outbursts One of the relentless and super motivated volunteer with the team is Mrs. Nalini Ramesh Shah, we thank her big enough for generously bestowing her time and talent. The efforts and the trust that she have been incessantly driving by rendering her charitable service in the field of non-invasive therapy using LILW from Medacsis to all those who are in dire needs.

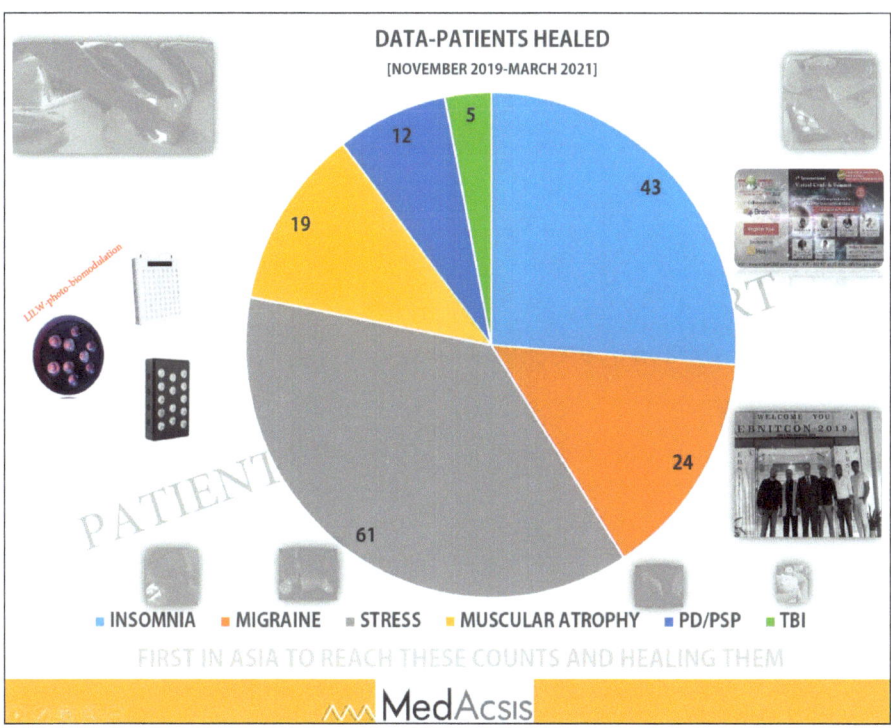

Light is not only the governor of hormones but the universe, thus by abusing light in whatever way, we imperil our health & physiology.

Just like, the ghrelin & leptin ways, we need to understand that how much our body demands for the light, when, where, how and which type.

Chapter 6

SLEEP

The functions of sleep:
1. Adaptive response
2. Restoration and repair of degenerative cells and neuronal activities
3. Adjusting the metabolism and strengthening the skeletal muscle.

Before we understand the dimensions and criterion of sleep and wakefulness, it is essential to know about Autonomic nervous system.

Autonomic nervous system is a black box of our body

Sleep & ANS are inter related as ANS is responsible sleep homeostasis. Regulates physiological of sleep stage

ANS is the coordinator of incoming and outgoing signals & assesses physiological traffic.

It coordinates so that the overall system is coduisive for the biological functions.

Sleep wake cycle is driven by the two processes:
1. Process c- indogeneous process, from SCN which are dependent on zeitgebers
2. Process s - the homeostatic drive, which relates to the amount and density of the sleep prior to the wakefulness. Control.

Control of parasympathetic & sympathetic pre ganglion are independent of circadian pathways from the level of hypothalamus to the brain stem. Circadian system is able to modulate both ANS sub division independently & balance the activity according to the time of day.

Some of the lighting organizations, individuals whether manufacturers or sellers or the consultants claims that the tunable white aligns the circadian processes, this is a myth. The correctness here I will like make is, it's just not only the SPD (Spectral power distribution whether R9 values or any other alterations), the introduction of bio red, bio blue and correctness of 585-595 nm are equally important to trigger the Process C of circadian rhythm or the melanopic pattern.

The immediate of sleepiness and fatigue:
1. Issues with the information processing and short term memory
2. Decreased performance & vigilance
3. Lower motivation
4. Erratic and aggressive behavior
5. Poor logics and creativity

The impact of chronic sleep disruption and reduced sleep on the promotion and interaction of physiological stress via the hypothalamic-pituitary-adrenal (HPA) and sympatho-adreno-medullary (SAM) axes and psychosocial stress.

Whereby sleep loss and fatigue result in an imbalance between the demands placed upon an individual and an inability of the individual to manage these demands.

Ultimately, the combined and interlocking effects of physiological and psychosocial stress lead to emotional, cognitive and physiological pathologies.

One of the sleep pattern refers to non-24-hour sleep, people who basically have no sleep problem but whose circadian clock is not entrained. Some blind people are not entrained to the 24-hour day despite a regular life involving work, meals and exercise. To understand why only some and not all blind people suffer from this problem, we must know many other important series of information.

The mammalian clock can only be influenced by light via the retina in the eyes. Those blind people, who have no conscious visual perception may still have a functional Melanopsin System that allows their circadian clock in the SCN to entrain to the 24-hour day.

Only those blind people, who have also lost light reception in the Melanopsin System are prone to be un-entrained despite a normal regular life.

- **Sleep changes with the age**
- Increased number and duration of awakenings
- Decreased REM sleep
- Decreased slow wave sleep
- Reduced sleep efficiency

- **Consequences of living against the clock**

Emotional response	Cognitive response	Somatic response
Increased irritability	Poor performance	Short naps
Mood fluctuations	Inattentiveness	Poor sleep pattern
Anxiety	Poor memory	Pain sensations
Depressed mood	Reduction in recall of events	Sensations of cold
Anger and aggression	Poor decision making	Cancer
Increased impulsivity	Reduction in the creativity	Misalignment of metabolism
Decreased motor skills	Poor productivity	Obesity
Exhaustiveness	Reduced concentration power	Drowsiness

Data from Pritchett D and WULFF K

The impact of sleep deprivation on cognitive & limbic brain function has been established by many researchers and the doctors.

This is not shocking since the societal and medical ramifications of the deficits caused by sleep loss, which commonly include impaired sustained attention together with deficits in learning and memory.

Moreover, the implication of such a relationship becomes particularly germane in the context of interpersonal as well as professional interactions that are predicated on the basis of emotional judgments and decisions, especially when considering the on-going erosion of obtained sleep time across many age ranges

(National Sleep Foundation poll 2007: http://www.sleepfoundation.org/).

To the best of my knowledge, there have been few empirical reports investigating the impact of sleep loss on tasks of emotion processing, with the majority of studies utilizing subjective rating scales.

Using objective ratings tasks, several studies have now investigated the effects of sleep and sleep disruption on emotional processing. It is also being argued that the critical contribution made by facial recognition to optimal human behaviour is exemplified by conditions of abnormal affective processing.

| Emotional response | Cognitive response | Somatic response |

For example, neurological patients who suffer from an inability to recognize certain emotions due to lesions in key extended limbic or prefrontal regions display marked social dysfunction, being unable to detect and subsequently be guided by relevant affective cues disorders such as autism provide additional evidence for impaired social interaction due to abnormal face processing, particularly expressions carrying emotional significance.

However, unlike the disorders described above, which express affective dysregulation due to structural brain abnormalities, the impact of acute sleep deprivation is, presumably, largely functional in nature, indicating there are additional circumstances, beyond substantive pathology, which can result in affective recognition deficits.

Considering that (a) impairments imposed by sleep deprivation were observed for some but not all emotions, (b) these deficits only became evident as the emotion gradient increased to moderate intensity levels.

ref: SRS, SRF, Els van der Helm, MSc; Ninad Gujar, MSc; Matthew P. Walker, PhD Sleep and Neuroimaging Laboratory, Department of Psychology and Helen Wills Neuroscience Institute, University of California, Berkeley, CA

I have come across the fact and analysis during my interaction with many on the issues of sleep deprivations that, Sleep deprived people might develop metabolic impairment over the people with good and quality sleep. Further to this sleep deprivation might increase in muscle atrophy & sarcopenia (It is characterized by the degenerative loss of skeletal muscle mass, quality, and strength).

To maintain the muscle mass & metabolism, MPT (Muscle protein turn over) is essential which can be generated thru good quality sleep.

Since the sleep and circadian rhythm disruption or disorder [SCRD] is eventually coz of mislaignment of internal and actual wakeful sleep clock, Ghrelin & Leptin imbalance ratios are cropped up leading to obesity and typeII diabetes.

Just to make you aware:
- **Ghrelin** also known as "hunger hormone" is a hormone produced by enteroendocrine cells of the gastrointestinal tract, it increases food intake. Blood levels of ghrelin are highest before meals when hungry, returning to lower levels after mealtimes.
- **Leptin** is a hormone in the small intestine that helps to regulate energy balance by inhibiting hunger, it gives the feeling of full stomach and signals the brain of restricting any food intake.

How this work: If we wake up at 7:00AM in the morning and our routine is to have supper by 9:00 PM, now for some reasons our sleep wake cycles are disturbed (examples in today's precarious situation overtly because of COVID-19) i.e. after dinner we used to sleep by 11 PM, which now (in case) shifted to 2: 00 AM the ghrelin will be triggered and we will like to stuff our intestines and then the Leptin will trigger in the morning not letting us eat or there can be several other cases like this.

In short the prolonged wakefulness and disruption in the sleep imbalances the triggering of these hormones leading to obesity, behavioral issues, typeII diabetes, weakened skeletal muscle ad on.

Sleep involves in the fundamental changes in discreet neural network. These neural network plays an essential role in behavior and pathology of mental disorders.

Please be informed to the fullest *"that because of lack of sleep we might develop the disorders like insomnia, migraine, and Alzheimer's & many behavioral disorders."*

So it is advisable that come what we may, all may need to get the good quality sleep every night or every day (shift works employees).

Sleep is one of the best treatments/medicine we can gift it our body and help in many aspects to stay healthy, wise & free from all the disorders which might occur because of the sleep deprivation.

CONCLUSION

Our body is not a machine but an assembly of living organisms & parts, which have to be aligned with Mother Nature

Light and all other factors definitely affects us, **let's get this straight we** *can work like robots but definitely we are NOT ROBOTS)*

Individuals/organizations who are involved in the study of lighting effects on Humans and plants primarily, decoding these into practical applications must get support. The support is very essential from the architects, interior designers, PMCs, the customers and the purchase departments

However, this cannot be achieved by ignoring significant parts of the biological mechanism, and validating the outcomes. As we all are aware that there are no shortcuts on the ladders of success. We need to strengthen our capabilities, abilities to pursue, forbearance, persistence and dedication.

Height of accomplishments is goes hands in hands with the depth of conviction and determination.

Use light & not abuse it, let's help all the inhabitants of our planet to have healthy & prosperous lives. Excuses are the nails to build the house of failures in all the aspects.

Let's leave the legacy in form of education, as the legacy is not just money and the assets but the applied form of knowledge we possess.

This can be the best return gift for our generations to come, before we depart for the heavenly abode.

"Lighting field is no less than medical field." As LIGHT is the governor of the universe and hormones too.

REFERENCES AND CREDITS

SRS, USA SRBR, USA. IES, iPAN, THE PHYSOC-UK, DR.WUNSCH, MEDACSIS, WIKIPEDIA, CIBSE/SLL, LMU-GERMANY-ONLINE, Dr. Till, Dr. Merrow, U O Michigan, USA-ONLINE,

ABOUT THERALICHT™ LLP

An Organisation with the Utilitarian Methodology for:
- LIGHTING SYSTEMS/DESIGN
- LIGHTING AUDITS
- TRAININGS
- ONLINE COURSES
- LIGHTING ADVISORY TOWARDS HEALTH, PROSPERITY & WELLBEING

Having the experience of over two decades in the relevant fields, THERALICHT™LLP is launched by having an integrating scientific approach, medical research and literature on environmental health, behavioural factors, health outcomes and demographic risk factors that affect health with leading practices in building design, construction and management.

THERALICHT™LLP offers the services to deal with the combined and interlocking effects of lighting related to the physiological, psychology and psychosocial stress which leads to emotional, cognitive, and physiological pathologies

Launched in December 2020, THERALICHT™LLP is one of the foremost organisation with individuals recognised as the inventors for the initiative to renovate the respective spaces and provide healthy yet comfortable along with the aesthetically appealing lighting systems/parameters for the occupants in germane ways that benefit people to prosper.

- **Specialists in**
- OFFICE LIGHTING SYSTEMS
- GREENWALL LIGHTING SYSTEMS
- RESIDENTIAL LIGHTING SYSTEMS

girish@theralicht.com
www.theralicht.com
www.medacsis.in
www.thedsagroup.com

www.ingramcontent.com/pod-product-compliance
Lightning Source LLC
Chambersburg PA
CBHW041105180526
45172CB00001B/112